the Galley Chef

A Healthy Guide for Hearty Eats Onboard the Vessel

QSE PUBLISHING
2008

QSE Publishing
A Division of Regal Enterprises, Inc.
23112 NE 144th Street
Woodinville, WA 98077

First Edition 2008

The QSE Publishing Web Site Address is
http://www.QSEPublishing.com

Library of Congress Control Number: 2007941753
ISBN 978-0-9796244-0-7

Printed In The United States of America

♻ Text Printed on Recycled Paper

DS Graphics
120 Stedman Street
Lowell, MA 01851

Our mission at QSE Publishing is to celebrate and promote quality, safety and environmental awareness worldwide. For more information on our publications, visit our website at QSEPublishing.com.

To order additional publications, please contact us at: office@qsepublishing.com.

WE WANT YOUR FAVORITE HEALTHY AND HEARTY RECIPES!

If you have a favorite healthy and hearty recipe you would like to share in our next edition of the *Galley Chef*, please submit your recipe, name and maritime company you work for to: office@qsepublishing.com

Thank you!

Cover: *The Pacific Titan Galley, a Western Towboat Company Tug*
Back Cover: *Black Bean Soup (page 55)*

This book is dedicated to the men and women who make a living on the water in a safe and environmentally responsible way – May your Galleys be healthy, sustaining you on your way.

CONTENTS

INTRODUCTION

The Galley is the heart of the boat, just like the kitchen is the heart of the home; where we come together over a meal to nourish our bodies and our souls. Some of our best memories are centered around food. That is why it is extremely important that meal time be a positive and safe experience. As a cook, it is necessary that you become well acquainted with the proper method to handle and prepare meals onboard the vessel. It is our desire that the *"Galley Chef"* will help you on your way to providing the safest and most enjoyable eating environment for you and your crewmembers.

Body Building Blocks

OIL
Maintains Healthy Skin, Hair and Energy

Minerals
Necessary for Electrical and Chemical Reactions

Vitamins
Assists Metabolism and Protein Production

Carbohydrates
Provides Energy

Water
Keeps the Blood Flowing and the Body Moving

Protein
Provides the Material for Growth and Renewal

√ Water – 8x8 Rule

Drinking at least 8 eight-ounce glasses of pure water throughout the day keeps your body functioning properly. In addition to pure water, you can hydrate with herbal teas and juices.

√ Carbohydrates – Fill Me Up

Hearty meals of pasta, breads, potatoes, and fruit provide fuel the body needs.

√ Protein – Animal & Vegetable

Legumes, nuts, seeds, grains and green vegetables are sources of protein in addition to animal sources, such as: meat, chicken, fish and dairy products.

√ Vitamins – A Bold Palette

Vegetables and fruits with bold color have optimal vitamin content: dark green broccoli and leafy vegetables; bright red peppers, apples, and grapes; deep purple eggplant and blueberries; and, rich orange carrots, squashes and sweet potatoes.

√ Minerals – A Delicate Balance

Mushrooms, sea vegetables, legumes and grains provide a wonderful source of potassium, iodine, zinc and iron.

√ Oils

Olive oil is one of the healthiest and most versatile oils you can cook with. Olive oil contains antioxidants called polyphenols. Be sure to shop for extra virgin olive oil. Save the refined vegetable oil for deep-frying. It is best to avoid peanut oil due to the severity of peanut allergies.

Must Have Foods in the Galley

Some foods pack more of a nutritional punch than others. Here is a list of must have foods:

√ Collard Greens, Kale, Spinach and Chard

The darker the leafy green the better. These vegetables reign supreme in folate, B12, and B6 for healthy minds and hearts.

√ Concord Grapes and Blueberries

These blues are bursting with antioxidants, which gobble up free radicals. Keep these in mind when selecting juice.

> Antioxidants prevent or slow the oxidation process in the body, which can damage cells and promote disease and aging.

√ Salmon, Sardines and Ground Flaxseed

Rich in omega-3 fatty acids, these three boost heart and brain health.

√ Whole Grains and Brown Rice

Have you heard the popular phrase: "The whiter the bread, the sooner you're dead"? Keep this in mind when you choose your grains. The whiter the grain, the more it has been processed and the essential vitamins and nutrients lost.

Oatmeal is a good grain to have onboard. You can use it to make a hearty breakfast and then use it for cookie or biscuit recipes.

√ Chocolate

Chocolate has been discovered as a mood enhancing, energy pick-me-up and antioxidant favorite treat. Select chocolate with the highest cocoa content, which is usually the darkest chocolate for heart healthy benefits.

√ Green Tea

For heart health and possible fat metabolism, try a cup of green tea instead of coffee.

Substitute whole wheat flour for white flour whenever possible; baked goods will be heartier and healthier.

√ Nuts and Seeds

Nuts are rich in antioxidants, especially almonds and walnuts. Seeds provide a wealth of nutrients. Be sure to select unprocessed and unsalted nuts and seeds. Texture and nutritional values are compromised when salted and processed.

√ Spices and Herbs

Spices and herbs add flavor to your foods, so you can cut back on fat, sugar and salt. Herbs freeze extremely well. If it is difficult to buy fresh herbs and freeze, then stock up on dried herbs, but not too much because the average shelf life for herbs and ground spices is 1 year.

√ Hard Cheese

Parmesan cheese keeps extremely well in the refrigerator because of its low moisture content. Finely grated parmesan cheese adds a nice touch to garnish pasta dishes and soups.

A good staple for the galley is containerized soup stock (vegetable or chicken). More and more you can find containers of organic soup stock on traditional grocery store shelves. The nutrients are more robust and contain less salt.

√ Fresh Onions, Celery and Garlic

Onion, garlic and celery are common ingredients in many dishes and all three store well in the refrigerator.

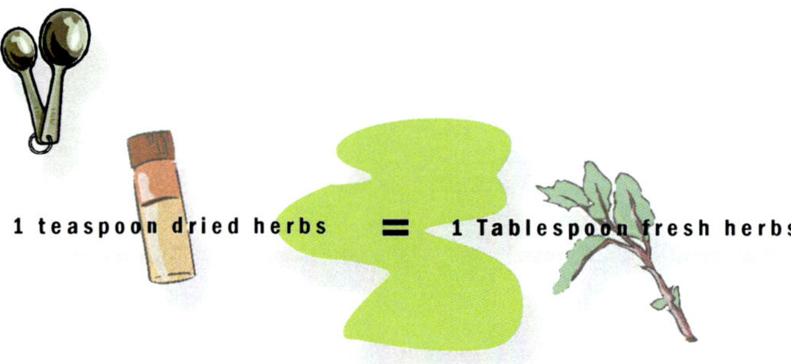

Try substituting: honey or maple syrup in your recipes for refined sugar; brown rice for white rice; and, organic for non-organic as much as possible. Avoid foods containing high fructose corn syrup and trans fats from hydrogenated oils.

READ LABELS CAREFULLY TO KNOW WHAT YOU ARE EATING!

Familiarize and Organize

It is important that you become familiar with everything you need as a cook and where it is located in the galley. If you find the galley is unorganized, set it up based on categories of items and frequency of use. Keep in mind that your personal preference of organization might not be the preference of the next person using the galley.

Use an inventory list to make sure you are adequately stocked before setting sail.

Basic Galley Tools

Using the right tool for the job, applies in the galley as well. Make your job easier by planning ahead and familiarizing yourself with the tools you have at your disposal. Below are some essential galley tools:

Bakeware and Bowls
Glass Baking Dishes (round, square and rectangular)
Jelly-Roll Pan (4 sided baking sheet)
Mixing Bowls
Muffin Pan
Wire Rack

Cutlery and Accessories
Cutting Board (plastic or other non-porous material)
Kitchen Scissors
Knife Sharpener
Sharp Knives

Equipment
Electric Mixer
Heavy Duty Blender
Pressure Cooker or Heavy Pot (for grains and beans)
Rice Cooker
Toaster/Toaster Oven

Gadgets and Utensils
Colander
Cooling Rack
Dry Measuring Cups and Spoons
Kitchen Thermometer
Ladle
Liquid Measuring Cups
Openers (Can and Bottle)
Pastry Brush
Pitcher
Pot Holders
Rolling Pin
Salad Dressing Bottle (with side measurements)

Gadgets and Utensils (continued)
Salad Spinner
Slotted Spoon
Spatulas
Strainer
Timer
Tongs
Vegetable Peeler
Vegetable Steamer
Versatile Grater (with different sides or attachments)
Wire Whisk
Wooden Spoons

Grocery Items
Aluminum Foil
Freezer Lock-Top Plastic Bags
Freezer Tape
Matches
Paper Towels
Permanent Marker for Labeling
Plastic Wrap
Soap
Toothpicks
Wax Paper

Pots and Pans
Dutch Oven with Lid
Frying Pans (large and small)
Large Soup Pot
Saucepans with Lids (small, medium and large)

Safety
First Aid Kit
Gloves

Food Storage

Storing food properly is the first step in handling food safely. Once provisions are brought aboard the vessel, items that need to be frozen or refrigerated should immediately be put away.

√ Pre-cool Fridge and Freezer

Check to ensure that the refrigerator is cooling between 32° to 40°F and that your freezer is cooling between -5° to +5°F. If the temperatures are off, you may need to adjust the dials on the unit to attain the right temperature. If you notice that the refrigerator or freezer is consistently too warm despite efforts to adjust the temperature, you should inform the captain because this represents a fairly serious danger.

√ Meat and Poultry

Any and all meat and poultry should be frozen if it's going to be more than 2 days before it's eaten (this includes sandwich meats).

Keep raw or uncooked meats separate from the rest of the items. Stow raw or uncooked meats below all other items in the fridge or freezer to ensure that these items do not drip or leak onto any other foods. Check the packaging to make sure that there aren't any rips or tears so that drips and leaks are prevented. If there are damaged packages, place inside a plastic bag or wrap in plastic wrap to create a leak-proof package.

√ Dairy and Eggs

Dairy items such as: milk, cheese, yogurt, and eggs should be stowed in the refrigerator. Ice cream, butter and any excess cheese should be frozen. Be sure to check your expiration dates on eggs and milk products.

Soft cheese (for example: mozzarella and havarti) freeze better than hard cheese like cheddar.

√ Fruits, Vegetables and Herbs

Produce items such as: lettuce, cauliflower, green onions, strawberries, fresh herbs, asparagus, or any precut or partially used produce, should be refrigerated. Other produce items should be stowed in a cool, dry and dark place onboard the vessel.

> When bananas get ripe faster than you can eat them, place in the freezer for fruit smoothies. (See recipe on page 50.)

√ Staples

Canned foods and dry goods should be stowed in cool, dry places. Check packages and expiration dates and make sure that the food is properly wrapped, sealed and fresh. Any damaged items, broken seals, or unlabelled cans indicate that the food should be disposed of. Once a dry good is opened, read the label to see if the item needs to be refrigerated after opening. Most condiments and salad dressings should be stowed in the refrigerator after the seal is broken.

> Bulk items should be transferred into easy to use containers.

√ Rotate Stock

Remember to rotate your stock. Stow by date; place oldest items in front, to be used first. Spend some time each day looking over your stores and produce for moldy, rotten, or past due food items. "When in doubt, throw it out."

Do not use canned goods that are dented, cracked or bulging!

SANITATION

Safety is first priority. If you are cooking onboard the vessel, you need to be as safety conscious as if you were handling lines or boarding a barge. The risk of food borne illness to the crew is as much of a threat as any other danger aboard the boat. It is necessary for you to be aware of the dangers of food borne illness, personal hygiene and the proper way to handle food safely for your own health and the health of everyone on board.

√ Food Borne Illness

Food borne illness often presents itself as flu-like symptoms such as: abdominal pain, nausea, vomiting, diarrhea or fever. Symptoms may occur as soon as one hour or as late as two weeks after eating spoiled food.

Bacteria may be present on food items when you purchase them like chicken, eggs, and certain vegetables and melons. Improper food handling, unsanitary conditions or poor personal hygiene can contaminate safe foods by introducing bacteria or other pathogens resulting in food borne illness.

√ Personal Hygiene Guidelines

- Wash hands frequently
- Wear clean clothes and apron
- Avoid wearing jewelry, false nails or other items that might fall into food
- Do not cough or sneeze over food
- Do not smoke around food
- Do not handle or prepare food if you are sick
- Keep fingernails trimmed short and filed
- Wear a smooth wedding band or none at all

√ Washing Your Hands

Washing your hands frequently is the single most effective thing you can do to prevent food borne illness. When you wash your hands, scrub with soap for at least 20 seconds and rinse with warm water, paying close attention to cleaning under fingernails and between fingers.

Wash your hands immediately before and after handling cooked and uncooked food, especially raw meat or poultry. Also wash your hands after:

- Going to the head
- Blowing your nose, sneezing or coughing
- Scratching your skin
- Touching your face or hair
- Smoking or eating
- Coming in contact with a any unclean surface

Food poisoning is commonly caused when someone handling the food wasn't diligent about keeping their hands as clean as possible.

√ Cover a Cut

Wear a single-use glove or finger cot to cover an open wound or infection.

Wearing gloves is an extra safety precaution you can take to lessen the chances of spreading germs and contaminating food. However, wearing gloves doesn't excuse you from keeping your hands clean. If you choose to wear gloves while handling food, you will need to change them after each time they have become contaminated.

√ Wear a Hair Net

If your hair is long, tie hair back and use a hair net or cap.

√ Clean as You Go

Adopt a "clean as you go" attitude and your galley will be a much safer and healthier place.

Food Prep Areas

When preparing food of any kind, make sure that the area in which you are using is clean and sanitary before you start. This includes making sure work surfaces, cutting boards, dishes, bowls and utensils are clean.

Wash cutting boards, dishes, utensils and counter tops with hot soapy water after preparing each food item and before you go on to the next food.

Counter tops and other work surfaces can be cleaned using a bleach and water solution (1 teaspoon of bleach to 1 quart of water) or any household "all-purpose" or "disinfectant" product (read the labels to make sure it is appropriate for counter-tops).

Food Storage Areas

While it is important to ensure that work surfaces and utensils are clean and sanitized, it is also important to make sure that areas where food is stored are clean and sanitized. This includes the refrigerator, freezer, forepeak, and cupboards.

When you are sanitizing any of these areas, remove the food from the area before using any soap or chemicals.

Food Preparation

√ Thawing Foods

Frozen foods frequently need to be thawed before cooking. There are three safe methods of thawing foods:

Refrigerator Method (18-36 hours)

The safest, but slowest method is to put the frozen item in the refrigerator until it has thawed. If you plan ahead properly, this should be the method you use at all times. When thawing in the refrigerator, remember to keep meat, poultry and fish sealed tightly and placed on the bottom shelf. Most items onboard the boat can be thawed in the refrigerator in 18-36 hours.

Cold Tap Water Method (4-6 hours)

Another method of thawing food is to submerge tightly wrapped food in cold tap water. This should be done in the sink or an area that will contain any leak or spill, such as a bowl. The cold tap water must be changed every 30 minutes while the item is thawing. This method typically takes between 4-6 hours for most items, although larger items may take considerably more time. It is important to remember that even though raw meat, poultry and fish is tightly wrapped, the water it is thawing in may still contain juices from the item. You should treat the water it is thawing in the same way you would treat touching the raw meat, poultry, or fish itself. Make sure all items are cleaned and sanitized properly before and after contact.

Microwave Method (Minutes)

The last method of properly thawing frozen foods is to use the microwave. Most microwaves have a "defrost" function that will thaw but avoid cooking the food. If you use this method to thaw, make sure the item is placed on a dish or container that will catch any spills or leaks from the frozen food. Also make sure that any packaging is microwave safe. If it is not, remove the packaging before thawing. After thawing, be sure to disinfect the inside of the microwave.

After an item is thawed, it should be cooked immediately. If an item is thawed and you are not ready to cook it yet, wrap it and return the food item to the refrigerator (it should be cooked within a few hours at most).

You may refreeze any uncooked item that was thawed using the refrigerator method, but if thawed by any other method, the item should be cooked before it is refrozen.

√ Separate for Safety

Keeping raw meats, poultry, fish, and their juices separate from other food items is very important. Not only should these things be kept separate from all other foods, they should be kept separate from each other.

Cross contamination occurs when raw or uncooked meats, poultry, or fish come in contact with each other. Cross contamination presents a high risk of causing food borne illness. If you are working with raw or uncooked meats, poultry, or fish you need to clean and sanitize all utensils, cutting boards, and work surfaces that were exposed during preparation before moving on to anything else. If you are preparing multiple types of meat, poultry, or fish, you need to clean and disinfect after handling each item before moving on to the next raw item. This includes washing your hands before and after handling each item.

If you are going to marinate any raw or uncooked meat, poultry, or fish, you should marinate it in a covered dish in the refrigerator on the bottom shelf.

Cooking

When it comes to preparing food, the greatest concern is that the food is handled and cooked properly, especially meat, poultry and fish. Since bacteria and germs are present, these items must be cooked until they are hot enough to kill the germs and bacteria. A thermometer is an important part of determining when the proper temperature has been reached. To use properly, insert the thermometer into the meat until the tip is in the center of the item and not touching any bone. Leave it there until the thermometer reads a constant temperature.

Using a food thermometer is the only safe way of knowing that the food has reached the right temperature to destroy harmful bacteria.

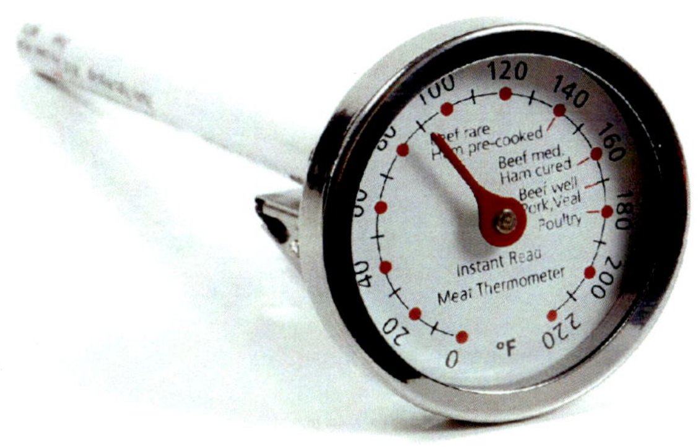

To be safe, cook stuffing outside the bird.

The following temperature guidelines refer to internal temperatures of properly cooked meat:

Poultry		170° - 180°F
All cuts of Pork	All cuts of Pork	160° - 170°F
Beef, Veal, Lamb	Medium Rare	145°F
	Medium	160°F
	Medium Well	160° – 170°F
	Well	170°F
Ground Meat	Ground Meat	160°F

Based on USDA Recommendations

Bacteria Multiply Quickly Between 40°F and 140°F

SERVING

When you are ready to serve the food you need to ensure that hot foods stay hot and cold foods stay cold. This is particularly the case when you are serving anything buffet style, where the food may sit for some time before it is eaten.

Hot foods should be kept at 140°F or higher and cold foods should be kept at 40°F or lower to prevent bacteria from forming in the food.

As a guideline, all cooked food should be refrigerated within 2 hours of cooking it. If a crewmember hasn't eaten within two hours of a meal being served, go ahead and refrigerate the food and he or she can reheat it when they are ready to eat. **Any cooked food left out for more than 2 hours should be discarded.**

√ Set the Stage

Nothing brings a group of people together like food. Presentation is very important when serving the meal. Having a table that looks full of inviting food can boost anyone's morale.

Celebrate holidays, birthdays or a day of the week with special food to create a festive environment. Roast a turkey with stuffing for Thanksgiving, bake cupcakes with frosting for a crewmember's birthday, make Saturdays Mexican Fiesta night or surf and turf.

Forks Go Left and Knives Go Right

If seas are calm, set the table the correct way. If you are underway and rolling around then buffet style is best.

Condiments and Extras

Take some extra time to place a variety of condiments on the table that go with the dishes served. Include small bowls of cut up raw vegetables and fruits as well as nuts for healthy side dishes. What isn't eaten can easily be stored for the next meal or snack.

Table Rules

Meal times are one of the few times the crew gathers as a community. It is important that all members be treated with respect and provided with a welcoming atmosphere at the table. Some common table rules are:

- Come to the table clean! No dirty, oily hands; no coveralls or sweaty work out clothes.
- Don't reach across someone to grab an item...ask.
- No obnoxious noises.

Be in Sync

Your galley table may serve just as a "fueling stop" for crewman about to work on deck; it may be the foundation of a "dinner and a movie night"; a lively discussion table; or, simply a place to "just eat in peace".

Vessel activities and crew's moods vary from day to day. As much as possible, your meal and presentation should bring out the positive aspects of the day and be in synch with the vessel's activities.

Leftovers and Perishable Items

Any food that will be kept as leftovers should be placed in relatively shallow containers and refrigerated. You will be in charge of making sure leftovers do not sit in the refrigerator for weeks on end. After 3 or 4 days in the refrigerator, discard any leftovers. If you have large quantities of leftovers, freeze and date stamp; use within a month or discard.

Be creative with leftovers by turning a rib roast into Philly steak sandwiches; a roasted chicken into Chicken Salad; and, a pork loin into pulled BBQ pork sandwiches. It is important to skip a day with leftovers to keep meals interesting.

Add leftover vegetables and food items that are close to being dated to soups and stews.

If you have many leftovers accumulating in the refrigerator, decide what can be placed in the freezer for future meals and what to bring out for a pot luck meal. The crew will most likely appreciate the variety if well presented.

Fruit salad is an excellent way to use up fruit before it becomes overly ripe. It is amazing how much fruit someone will eat if cut up and presented as a salad or side dish.

Meal Planning

Paying attention to nutrition and planning out your meals in advance combined with proper food handling techniques are key parts of the cook's job. Other crew members rely on you to know and understand how important your job is to their health and well-being while onboard the vessel.

√ The Food Balancing Act

Preparing a balanced diet of lean proteins, fresh fruits, whole grains, nuts and vegetables gives the crew the energy they need to maintain peak performance. A balanced diet also helps the crew to feel better physically and psychologically.

Every meal should contain a source of protein (the leaner the better), carbohydrate and vegetable. Your goal should be to prepare balanced meals that incorporate enough calories and nutrients for everyone.

Calories

Food provides energy to fuel your body. Energy is measured as the number of calories a food contains (the more calories, the greater amount of fuel the food provides). On average, an adult, relatively active man will burn about 2500 calories a day; 1800 for women. If you consume more calories than you burn it turns to fat. Calorie content in foods is almost always provided on the food label. It would be difficult and time consuming to measure the calorie content of each meal. However, it is good to have a working knowledge of what food items are high and which are low in calories.

Healthy Snacks

Many times crew members eat on the run, so snacking is common and necessary. By having a variety of healthy alternatives available to the crew you will help them to avoid high caloric and nutritionally void "filler" food such as chips and candy bars. Here are some suggestions of healthy snack items you can provide the crew:

- Sliced raw fruits and vegetables
- Fresh baked goods
- Granola
- Sliced or cubed cheese
- Nut mix with dried fruit
- Small sandwiches like roll-ups

Caffeine

Life onboard workboats is usually a 24/7 operation; it is common to have a coffee pot brewing around the clock. Caffeine has its place as a stimulant to boost alertness when necessary but it should be managed carefully.

Besides its stimulating effect, caffeine can also have some detrimental secondary effects. What you may not know is that high doses of caffeine consumption can result in increased anxiety, heart palpitations and digestive problems. Probably the most significant detriment aboard vessels is that caffeine can also interfere with natural sleep patterns, keeping you awake when you want to sleep.

Managing rest cycles is a priority onboard any workboat. A rest period may come at irregular times or be shortened by changes in the vessel's operations schedule. Because of caffeine's powerful and lasting stimulant effect, its intake should be managed so as to not interfere with these valuable rest periods. The USCG recommends that caffeine be avoided altogether within 4-hours of a rest period.

Caffeine is not only present in coffee, but soft drinks, tea and medications as well. Sometimes a person can consume significant amounts of caffeine without knowing it. Be sure to read the label for caffeine content.

Beverage	Approximate Caffeine Content (mg)
Coffee (Drip)	130
Soft Drinks	40
Tea	30

If you need or want to limit your caffeine intake, try substituting out a cup of regular coffee with:

- Decaffeinated Coffee
- Green Tea
- Herbal Tea
- Sparkling Water
- 100% Vegetable Juice or Fruit Juice

Rules for a Healthy and Balanced Diet

- **Use Whole Foods:** Use fresh or frozen whole foods as much as possible instead of processed foods; the shorter the ingredients list the better.

- **Ensure Variety:** Ensure a broad range of nutrients by eating a variety of color; choose the deepest and brightest colors for optimal vitamin and mineral content.

- **Increase Plant Foods; Decrease Animal Fat:** Increase vegetables, fruits, whole grains, beans, fish and nuts and decrease the use of animal fat for all around body health.

√ Know Your Crew

Find out as soon as possible what your crew's preferences are: who eats breakfast, who doesn't; does the Captain have a special meal request; who has food intolerances or allergies and to what. Always ask what the crew likes to get their input. They will greatly appreciate you wanting to accommodate them as much as possible.

√ Food Intolerances or Allergies

Always check with crew members to find out if they have any food intolerances or allergies. Sometimes the ingredients you have included in a dish may not be entirely obvious to those eating. You want to be conscious of items that may cause allergies of any kind. If a crew member does have a food allergy, find out from them what symptoms they exhibit when coming in contact with that food, so you will know what to look for if there is a problem. Make sure to keep ingredients that someone is allergic to entirely separate from any foods they will be eating, including using separate utensils and cleaning work surfaces that come in contact with food that may trigger allergies.

Some people have quite violent and life-threatening reactions to specific foods, so their well being can quite literally be in your hands if you are preparing or cooking food for them.

> Know the signs of an allergic reaction, both minor and severe. Know how to treat them immediately.

√ Plan Ahead

Plan a day or two ahead for what you intend to cook at each meal. By planning ahead, literally sitting down and writing out what you will make at each meal, you can spend some time making sure meals are nutritional and that all the items you will need for the meal are properly thawed and ready when you begin cooking.

At the start of each trip, consult the captain as to when and what meals you are expected to prepare. After you know what responsibilities are expected of you, you can properly plan meals in advance. Planning will also help you get an idea about how long it will take you to prepare each meal before you start. Staying organized in your duties will make your work a lot easier.

> When determining how much food to make for each meal, take your crew base and add 2.

Rough Seas or Calm Waters

Always consider the weather when cooking onboard any vessel. Rough seas can make preparing certain foods far more difficult than in calm waters. Weather conditions can change rapidly; even if things are peaceful one moment, a few hours could result in a complete change of weather, making cooking much more difficult. This is another reason meal planning is a good practice. Consult the captain or mate about weather conditions a few hours before each meal so that you can allow yourself extra time if you need it.

In Port

Prior to arriving in port, make a casserole like lasagna the day before so it can easily be placed in the oven. Add salad and bread and you have an easy and satisfying meal that the crew will appreciate.

> Plan a few pre-made meals in advance that you can freeze for when the rough weather hits. A pre-made meal is a whole lot easier to get on the table in rough seas, when all you have to do is put it in the oven.

Measurements and Equivalents

1 Gallon	=	4 Quarts
1 Quart	=	4 Cups
1 Pint	=	2 Cups
1 Cup	=	16 Tablespoons
1 Tablespoon	=	3 Teaspoons
1 Pinch	=	Slightly less than ¼ Teaspoon

7-Day Sample Menu

Meals on workboats serve dual functions due to crew members' work schedules. As an example, the crew on the 12-6 watch will look at lunch as their main meal. They will need a menu that provides them with enough balanced calories to serve as "dinner". Additionally they will look at dinner as a "snack" before they go to sleep and should be provided with lighter calorie menu options.

MONDAY	
Breakfast	• Bacon and Eggs to Order • Hash Browns • Toast and Fruit • Juice, Milk, Coffee and Tea
Lunch	• Navy Bean and Ham Soup (page 59) • Sandwiches
Dinner	• Marinated Teriyaki Chicken Breast (page 107) with Teriyaki Dipping Sauce (page 106) • Steamed Jasmine Rice • Steamed Vegetables • Salad
Midnight Meal or Snack	• Soup • Sliced Fresh Fruit or Vegetables

TUESDAY	
Breakfast	• Oatmeal or Cereal • Fruit and Toast or Muffin • Yogurt • Juice, Milk, Coffee and Tea
Lunch	• Chicken Caesar Salad (page 75) • Dinner rolls or Bread Sticks
Dinner	• Spicy Apple Pork Loin (page 101) • Pork Fried Rice (page 97) • Apple Sauce • Steamed Vegetables • Salad
Midnight Meal or Snack	• Navy Bean and Ham Soup • Sliced Fresh Fruit and Vegetables

WEDNESDAY	
Breakfast	• Waffles, Link Sausage and Eggs to Order • Fruit • Juice, Milk, Coffee and Tea
Lunch	• Chicken Soft Tacos (page 79) • Mexican Red Rice and Beans (page 92)
Dinner	• Macadamia Nut Crusted Halibut (page 89) • Orzo Pasta with Shitake Mushrooms (page 93) • Steamed Vegetables • Salad
Midnight Meal or Snack	• Leftover Marinated Teriyaki Chicken Breast Sandwiches with Teriyaki Dipping Sauce • Sliced Fresh Fruit and Vegetables

THURSDAY	
Breakfast	• Oatmeal or Cereal • Fruit and Toast or Muffin • Yogurt • Juice, Milk, Coffee and Tea
Lunch	• Spicy Pot Sticker Soup (page 60) • Sandwiches
Dinner	• Pasta and Meatballs with Marinara Sauce (page 91) • Garlic Baguette Bread • Caesar Salad
Midnight Meal or Snack	• Leftover Spicy Apple Pork Loin Sandwiches • Sliced Fresh Fruit and Vegetables

FRIDAY	
Breakfast	• Poached Eggs on Toast and Ham Steak • Juice, Milk, Coffee and Tea
Lunch	• Black Bean Soup (page 55) with Corn Chips and Salsa
Dinner	• Marinated Steak (page 103)with Balsamic Vinegar Reduction Sauce (page 65) • Garlic Mashed Potatoes (page 87) • Vegetables • Salad
Midnight Meal or Snack	• Orzo Pasta Soup and Sandwiches • Sliced Fresh Fruit and Vegetables

SATURDAY	
Breakfast	• Pancakes, Sausage Patties and Eggs to Order • Fruit • Juice, Milk, Coffee and Tea
Lunch	• Halibut or White Fish Chowder (page 56) with Sheet Pan Bread (page 52)
Dinner	• Blackened Fish (page 71) with Creole Mustard Sauce (page 81) • Boiled Baby Red Potatoes with Dill • Steamed Vegetables • Salad
Midnight Meal or Snack	• Spicy Pot Sticker Soup • Sliced Fresh Fruit and Vegetables

SUNDAY	
Breakfast	• French Toast Casserole (page), Sausage Patties • Juice, Milk, Coffee and Tea
Lunch	• Seared Ceaser Halibut Cheeks (page 96) on Kaiser Roll • Carrot Fries (page 69)
Dinner	• Beef Burgundy (page 66) with Rice or Wide Egg Noodles • Steamed Vegetables • Dinner Rolls • Salad
Midnight Meal or Snack	• Black Bean Soup • Sliced Fresh Fruit and Vegetables

My Favorite 7-Day Menu

MONDAY	
Breakfast	
Lunch	
Dinner	
Midnight Meal or Snack	

TUESDAY	
Breakfast	
Lunch	
Dinner	
Midnight Meal or Snack	

WEDNESDAY	
Breakfast	
Lunch	
Dinner	
Midnight Meal or Snack	

THURSDAY	
Breakfast	
Lunch	
Dinner	
Midnight Meal or Snack	

FRIDAY	
Breakfast	
Lunch	
Dinner	
Midnight Meal or Snack	

SATURDAY	
Breakfast	
Lunch	
Dinner	
Midnight Meal or Snack	

SUNDAY	
Breakfast	
Lunch	
Dinner	
Midnight Meal or Snack	

Healthy and Hearty Recipes

Amish Baked Oatmeal

Courtesy of Matthew Broomhead, Sea Coast Transportation

Serves: 4

1 ½	cups rolled oats
½	cup sugar
1	teaspoon baking powder
½	teaspoon salt
1	teaspoon ground cinnamon
½	cup milk
1	egg
¼	cup melted butter
1	teaspoon vanilla extract

1. Preheat oven to 350°.
2. In a large bowl, mix together oats, sugar, baking powder, salt and cinnamon.
3. Beat in milk, egg, melted butter, and vanilla extract.
4. Spread into a greased 9x13 inch baking dish.
5. Bake for 25-30 minutes or until edges are golden brown.
6. Immediately spoon into bowls.

Serving Suggestion: Serve with warm milk and brown sugar.

Don't forget to top your yogurt or cereal with nutritios food like, nuts, seeds, ground flax and fruit (fresh or dried).

Oatmeal Blueberry Muffins

Courtesy of Sara Mazurek, Crowley Maritime Corporation

Serves: 4-5

1	egg
1	cup buttermilk or yogurt as substitute
⅓	cup softened butter
½	cup brown sugar
1	cup rolled oats
½	cup whole wheat flour
½	cup white flour
1	teaspoon baking powder
1	teaspoon salt
½	teaspoon baking soda
1	cup blueberries or other fruit

1. Preheat oven to 400°F.
2. Spray or wipe muffin tin with oil, butter or non-stick spray.
3. Mix wet ingredients together.
4. In a separate larger bowl, mix dry ingredients together.
5. Fold wet into dry ingredients, be careful not to over mix.
6. Add fresh or frozen blueberries.
7. Pour into muffin tin and bake until golden brown, about 15-20 minutes. Insert toothpick in center to see if they are ready; the toothpick will come out clean.
8. Remove tin from oven, let sit for about 5 minutes to cool and then remove muffins.

French Toast Casserole

Courtesy of Chris W. Starkenburg, Harley Marine Services

6	Tablespoons of butter at room temperature
12	pieces thick slice egg bread (Texas Toast)
12	whole eggs
3½	cups half-and-half or milk of choice
8	ounces cream cheese at room temperature
8	ounces raspberry jam or favorite preserve
1	Tablespoon cinnamon
1	Tablespoon vanilla extract
	real maple syrup

1. Completely coat the bottom and both sides of a glass, stainless or non-stick baking dish 13"x9"x2" with butter.
2. Combine eggs, milk or half-and-half, cinnamon and vanilla in a mixing bowl and whisk until blended; set-aside.
3. Lay out your slices of egg bread on counter and spread half the slices with cream cheese and the other half with jam, match them together like a sandwich.
4. With a bread knife, cube the sandwich into 1"x1"cubes.
5. Place cubes into buttered baking dish.
6. Gently pour egg mixture over cubed bread.
7. Place a piece of plastic wrap over the top and gently press down so that the cubed bread gets coated with egg mixture.
8. Place in refrigerator overnight or for at least 6 hours.
9. Remove plastic wrap and place in preheated 325°F oven and bake for 45-50 minutes. Cooking time may vary; when you insert a knife it should come out clean.
10. Let cool for 10 minutes; slice and top with maple syrup or your favorite topping.

Fruit Smoothies

Serves: 1-2

1½	cups juice
2	partially frozen bananas with skin removed

Place ingredients in a blender and puree until smooth.

Waffles and pancakes freeze and reheat in the toaster extremely well. If making waffles, double the batch and freeze the rest for later.

Pork Sausage and Eggs

Serves: 5-6

1	pound of pork sausage
9	eggs, slightly beaten
3	cups milk
1½	teaspoon dry mustard
3	slices of bread, cubed
1½	cups cheddar cheese, grated

1. Brown meat, drain and cut meat into bite size pieces after cooking.
2. Mix eggs, milk and seasonings.
3. Stir in bread, cheese and sausage.
4. Pour in 9"x13" pan.
5. Refrigerate overnight covered.
6. Bake uncovered at 350°F for 1 hour.

Stuffed French Toast

Courtesy of Chris W. Starkenburg, Harley Marine Services

Serves: 6-7

6	whole eggs
1½	cup milk or half-and-half
1	Tablespoon pure vanilla
4-6	ounces cream cheese
1	loaf thick sliced egg bread
6	whole strawberries, thinly sliced or other fresh fruit
	strawberry jam or favorite preserve
	Hot Cinnamon Syrup:
1	cup butter
2	cups real maple syrup
2	Tablespoons ground cinnamon

1. Make Hot Cinnamon Syrup by placing butter, maple syrup and 1½ Tablespoons of cinnamon in a sauce pan on medium to medium low heat.
2. Stir until completely blended and warm, set aside.
3. In large bowl, whisk eggs, milk or half-and-half and vanilla until completely blended.
4. Warm cream cheese in microwave until it is easy to spread.
5. Spread the cream cheese on one side of the egg bread.
6. Place a thin layer of strawberries or your favorite fruit on top.
7. Generously coat another piece of the egg bread with jam or your favorite preserve, and sandwich together with the cream cheese side.
8. Preheat a non stick skillet on medium high heat.
9. Place a little butter in skillet; dip both sides of the stuffed french toast in the egg wash mixture; then place in the skillet.
10. Sprinkle the top with cinnamon.
11. Cook until golden brown, and then repeat the process on the other side.
12. Remove from pan, place on a plate and top with Hot Cinnamon Syrup.

Sheet Pan Bread

Courtesy of Chris W. Starkenburg, Harley Marine Services

Serves: 6-8

2½	cups warm water (105-110°F)
2	packages yeast
3	Tablespoons flour
1	Tablespoon sugar
¼	cup olive oil
1	teaspoon kosher salt
4¾	cups flour

1. Place the first 4 ingredients in a mixing bowl and blend slowly on low speed for about 1 minute.
2. Cover the bowl with plastic wrap and a clean unscented towel.
3. After the mixture starts to foam (approximately 10-15 minutes), add the last 3 ingredients: olive oil, salt and flour to your bowl.
4. Mix with a mixer (add dough hook) or by hand, until all ingredients are blended (approximately 10 minutes).
5. Scrap the dough onto a well floured board or counter. The dough will be sticky and not come out of the bowl easily.
6. Knead the dough for 5 minutes, adding more flour as necessary to form a nice soft, velvety ball.
7. Place the dough in a lightly oiled bowl or place in the center of your board and cover with the plastic wrap towel.
8. Once the dough has doubled in size, approximately 10-15 minutes, punch down and knead once more.
9. With a floured rolling pin, roll the dough until it is the size of the sheet pan you will be baking it in.
10. Place dough onto a greased (use olive oil) sheet pan and spread evenly to cover the pan.

11. Rub the top of the dough with olive oil and sprinkle a generous amount of salt and your favorite herb (optional).
12. Preheat oven to 350°F.
13. Let sit at room temperature for 15 minutes.
14. Place in oven and bake until golden brown (approximately 20-25 minutes). For crispier bread, place the pan on an oven stone.
15. Remove from oven and let set for 5-8 minutes before cutting.

Serving Suggestion: If you don't have time to make just before dinner, make everything in the morning and place the sheet pan in the refrigerator until you are ready to bake. Just remember you will need to bring it up to room temperature (proof the dough) and let it start rising before you bake, which will take about 35-45 minutes depending on how warm your galley is.

Sweet Corn Bread, page 54

Sweet Corn Bread

Courtesy of Chris W. Starkenburg, Harley Marine Services

Serves: 6-8

½	cup of sweet butter cut in half, melt half for liquid mixture
3	cups all-purpose flour
1	cup granulated sugar
1½	cups yellow ground cornmeal
1½	Tablespoons baking powder
2	teaspoons kosher salt
1	cup milk or half-and-half
¾	cup honey
3	eggs, beaten
1½	cups canned cream corn
¾	cup vegetable oil

1. Preheat: oven to 350°F.
2. Add ¼ cup of butter to a large cast-iron skillet and place in oven.
3. Combine all dry ingredients: flour, sugar, cornmeal, baking powder and salt in a medium size bowl.
4. Mix ingredients.
5. In a separate bowl, mix all liquid ingredients until blended.
6. Add to dry ingredients and mix until blended.
7. When the butter in the skillet is melted and lightly brown (nut brown) pour in cornbread mixture.
8. Place skillet back in oven and cook (approximately 20-25 minutes, or until you place a knife down the center and it comes out clean).
9. Carefully remove from the oven and let cool for at least 10 minutes.

Black Bean Soup

Courtesy of Chris W. Starkenburg, Harley Marine Services

2	Tablespoons vegetable or olive oil	2½	Tablespoons chili powder
1	cup onion, chopped	2	Tablespoons parsley
½	cup celery, chopped	½	teaspoon crushed red chilies
½	cup carrots, diced	1	Tablespoon Tabasco sauce
1½	Tablespoons garlic, chopped	4	Tablespoons chicken base (or 8 cups chicken broth and omit water below)
1	teaspoon whole thyme	8	cups hot water
1	teaspoon black pepper	6	cans black beans
2	Tablespoons ground cumin		

1. In a soup pot, sauté onions, celery, carrots and garlic in oil until tender.
2. Add herbs and spices, sautéing on low heat for 10 minutes.
3. Dissolve chicken base in water or use prepared chicken broth and add to sautéed mixture.
4. Puree 3 cans of beans with juice in blender, add to mixture.
5. Add last 3 cans of beans, drained well but not rinsed.
6. Simmer at low heat for 1 hour.

Serving Suggestion: Top with sour cream and serve with tortilla chips.

Halibut or White Fish Chowder

Courtesy of Chris W. Starkenburg, Harley Marine Services

Serves: 10-12

2½ -3	pounds boneless and skinless halibut or white fish filets	½	teaspoon chili powder
3	cups water	1	teaspoon ground white pepper
1	pound bacon, diced	2	pinches cayenne pepper
2	cups celery, chopped	3-4	Tablespoons of clam base (or bottled clam juice to taste and omit water below)
1½	cups onion, diced	½	cup hot water
1	cup butter	6-7	cups of half-and-half or milk
2	cups white flour	10	cups red potatoes, diced, parboiled firm but tender, and drained but do not rinse
¾	Tablespoon whole thyme		green onions (for garnish)

1. Place filets in a large baking dish and cover with 3 cups of water.
2. Cover the dish with foil and place in a 350°F oven.
3. Poach fish for about 10-12 minutes or until filets are cooked through.
4. Remove from the oven, drain off the liquid (save liquid), and place filets in the refrigerator to stop the cooking process.
5. In a large pot, cook the diced bacon until medium-well done on low to medium heat, stirring consistently as to not let the bacon brown.
6. Add celery, onion and butter and cook until tender or until onions are transparent. Then add the flour and spices.
7. Reduce heat and stir constantly for at least 12 minutes, do not brown mixture (roux).
8. Dissolve clam base in hot water or add bottle clam juice to taste.
9. Stir in all the liquids, including the poaching liquid and let the stock thicken.

10. Add the potatoes and continue to stir, do not let the bottom scorch. Once the stock is to desired thickness, transfer the pot into a double boiler or crock pot (on low) to keep warm and prevent from burning.
11. About 20-30 minutes before serving, remove fish filets from your refrigerator and break into 1"x1" chunks. Gently fold the fish into your chowder base, and continue to warm. Remember the more you stir the smaller the fish chunks will get. This is a good time to add other types of seafood that is cooked.

Cooking Suggestion: Keep the chowder in a double boiler to prevent the chowder from scorching. You can also transfer the finished product to a Crock Pot, but keep the temperature setting on low.

Serving Suggestion: Place chowder in bowls and garnish with crispy bacon bits and thinly sliced green onions.

Egg Substitution
1 egg = 1/3 cup water + 4 teaspoons ground flax seed boiled for 1 minute to thicken

Minestrone Soup

Serves: 4-6

4	Tablespoons olive oil	1	teaspoon dried basil
4	cloves garlic, crushed	1	cup green, orange or yellow pepper, chopped
1	cup onion, chopped	4	cups vegetable stock or water
1	teaspoon salt	2	(14.5 ounce) cans stewed tomatoes
1	cup celery, chopped	1	(15 ounce) can kidney beans
1	cup carrot, sliced	2	cups fresh or frozen corn
1	pinch of black pepper	1	cup summer squash (zucchini) if available, cubed
1	teaspoon dried oregano	1	cup uncooked pasta
1	teaspoon dried parsley		salt and pepper to taste

1. In a large soup pot, sauté garlic and onions in olive oil on medium heat until soft and translucent.
2. Add salt, celery and carrots, mixing well.
3. Add herbs and stir.
4. Cover and cook over low heat for 5 minutes.
5. Add peppers, vegetable stock and undrained tomatoes.
6. Bring to a boil and reduce heat. Simmer for about 15 minutes.
7. Stir in the beans, corn, summer squash (if available) and pasta; simmer for 10 to 15 more minutes until the vegetables are tender. Season with salt and pepper.

Serving Suggestion: Garnish with grated parmesan cheese and serve with bread.

Minestrone soup freezes well. Double the recipe for a quick meal when the weather gets rough.

Navy Bean and Ham Soup

Courtesy of Chris W. Starkenburg, Harley Marine Services

Serves: 6-8

32	ounce bag small white beans
13	cups water
3-4	Tablespoons soup base
1	bay leaf
¼	teaspoon black pepper
1	Tablespoon parsley flakes
1½	cups celery, chopped
1	cup yellow onion, chopped
1	cup carrot, grated or diced
3	cups ham or boiled down smoked ham hocks pulled from bone, diced
	salt and pepper to taste

1. Soak beans for at least 4 hours in cold water. Always keep beans covered in water and leave ample room for expansion; or, you can soak them overnight.
2. Drain and place in large soup pot.
3. Add 13 cups of water, soup base, bay leaf and pepper.
4. Bring to a boil then reduce heat and simmer until beans are tender, about 1 hour and 15 minutes.
5. Remove 2 cups of the beans and set aside.
6. Add the rest of the ingredients, and continue to simmer for another hour. Do not let the beans start to brake down.
7. Mash up the beans that you set aside and add back into the soup, this will help thicken the soup.
8. Add salt and pepper to taste.

Crock Pot Method: After soaking the beans, add everything to the crock pot and cook for 6-7 hours. When the beans are tender, remove 2 cups, mash and return to crock pot to help thicken soup.

Spicy Pot Sticker Soup

Courtesy of Chris W. Starkenburg, Harley Marine Services

Serves: 6-8

1½	Tablespoons sesame seed oil	1	can sliced water chestnuts
1	Tablespoon olive oil	1	can sliced bamboo shoots
1	medium size onion, julienne	1	can bean sprouts
4	stocks celery, chevron cut on bias (across the grain)	3	boneless chicken breasts , sautéed in separate pan then diced
3	carrots, pealed and julienne	4	Tablespoons chicken base (or 4-5 cups chicken broth and omit water below)
1	teaspoons crushed red chilies	4-5	cups hot water
2	teaspoons ground mustard (powder)	2	bags (24 each) frozen pot stickers (any flavor)
1½	teaspoons ground white pepper	1	cup green onions, chevron cut (for garnish)

1. In large soup pot on medium heat, add oils (sesame and olive), onion, celery, and carrots. Cook for about 4-5 minutes, stirring occasionally.
2. Add spices and the rest of the ingredients except the pot stickers and green onions.
3. Bring to a slow simmer and reduce heat to the lowest setting. Only keep warm for about 1 hour before serving so you don't over cook the vegetables.
4. In another 5-6 quart pot bring 4 quarts of water and ⅓ cup vegetable oil to a boil.
5. Add about 12 pot stickers at a time and cook for about 5-6 minutes, this is a little less time than the instructions.
6. Remove and place in an ice bath to stop the cooking process.
7. About 10 minutes before serving, add pot stickers back to soup and bring back to a simmer then remove pot from heat.

Serving Suggestion: Serve in a large bowl and garnish the top with green onions.

Storing Suggestion: To save for leftovers, remove the pot stickers so they don't break apart. Refrigerate the pot stickers and broth separately.

White Bean Chili

Courtesy of Chris W. Starkenburg, Harley Marine Services

Serves: 6-8

4	Tablespoons olive oil (omit if you use bacon)	1	Tablespoon Tabasco sauce
4	strips cooked bacon, diced (optional)		kosher salt to taste
4	stocks celery, diced	2	Tablespoons fresh basil, chopped fine
1	medium onion, chopped	1	Tablespoon fresh cilantro, chopped fine
1	small red bell pepper, diced	1	lime, the zest and the juice
2	large Tablespoons garlic, chopped	1	tomato, seeded and diced
1½	teaspoons crushed red chili flakes	4	cups cooked white chicken breast meat, chopped (charbroil the breast for extra flavor)
1	teaspoon white pepper	4	Tablespoons chicken base (or 4-5 cups chicken broth and omit water below)
1	Tablespoon chili powder	4-5	cups hot water
1	Tablespoon ground cumin	7	cans (15.5 ounce) great northern white beans, drained and rinsed

1. If you choose to use the bacon, place in soup pot and cook until crisp and remove.
2. If you don't use bacon, replace with olive oil.
3. Add celery, onion and pepper; sauté until tender, then add all the spices and cook for another minute.
4. Add all remaining ingredients including bacon bits and simmer for several hours on low heat until you are ready to eat.
5. Before serving, remove 2 cups of chili and puree in blender and add back into the pot, this will help thicken with out loosing flavor.

Crock Pot Method: Place all ingredients in the crock pot; stir until mixed; and, cook for 6 hours.

Serving Suggestion: Garnish with tortilla chips, green onions, drizzle of sour cream, black beans or crisp bacon bits.

Au Jus

Courtesy of Chris W. Starkenburg, Harley Marine Services

2½	cups Merlot wine
5	cloves of garlic
1	bay leaf
4-5	Tablespoons beef base
6	cups hot water
20	whole peppercorns
	Cornstarch Slurry:
1	Tablespoon cornstarch
2	Tablespoons cold water

1. Place all ingredients, except the Cornstarch Slurry in a saucepan and slowly simmer for 2-3 hours.
2. Taste the Au Jus for flavor; adjust with either more water or more beef base.
3. If the Au Jus is to your liking, bring it to a slow boil.
4. Stir in Cornstarch Slurry.
5. Continue to stir until Au Jus returns to a boil, then reduce the heat. Do not continue to boil or the cornstarch will breakdown.

Arroz Con Pollo

Courtesy of Matthew Broomhead, Sea Coast Transportation

2-4	pounds chicken breast halves or thighs without skin	4	drops Tabasco sauce
4-5	teaspoons Goya Seasoning, page 87	½	lime sqeezed for juice
2	Tablespoons olive oil	½	green pepper, cubed
2	cloves garlic, minced	3	Tablespoon green olives, sliced
½	onion, diced	1	Tablespoon capers
1¼	cans 16 ounce tomato sauce + ½ can water	½	teaspoon sugar
			Fresh colantro (for garnish)

1. Cut chicken into large cubes.
2. Rub Goya Seasoning over chicken.
3. Heat oil in a large skillet over medium high heat; sauté garlic and onions.
4. Add chicken; cook covered until golden on all sides, about 10 minutes.
5. Add tomato sauce with water, Tabasco, lime juice, green pepper, green olives, capers and sugar to skillet; stir, cover and simmer on medium heat until vegetables are tender, about 20 minutes.
6. Before serving, stir in small handful of coarsely chopped cilantro.

Serving Suggestion: Serve with Jasmine Rice and Frijoles (Beans).

Baked Acorn Squash with Bacon

Courtesy of Chris W. Starkenburg, Harley Marine Services

3	acorn squash
1	cup brown sugar
6-8	strips smoked bacon, cut in half
3	Tablespoons butter
2	Tablespoons maple syrup
1-2	cups water
	kosher salt and pepper to taste

1. Preheat oven to 350°F.
2. Cut Squash in half and remove seeds.
3. Place in baking dish flesh side down, add water to dish and bake for 35 minutes.
4. Remove from oven and remove half the water and turn squash skin side down.
5. Sprinkle with salt and pepper, then lightly sprinkle with brown sugar and place bacon on top of squash; place back in oven and continue to bake until squash will easily scoop from skin, about 30 minutes.
6. Remove from oven and discard the water in the bottom of the pan and let cool until you can handle.
7. Remove the bacon and dice into bits and if needed cook until crisp.
8. In a large bowl scoop the flesh from squash along with the liquid that was left in the squash, add butter and bacon bits; mix until smooth.
9. Taste for salt pepper and sweetness and then place in casserole dish.
10. Reheat just before serving and garnish with bacon bits and drizzle with maple syrup.

Cooking Suggestion: This can be made the day before.

Balsamic Vinegar Reduction Sauce

Courtesy of Chris W. Starkenburg, Harley Marine Services

1	cup balsamic vinegar
1	can beef broth
6	whole black peppercorns
1	fresh whole garlic clove

1. Place all ingredients in a stainless steel saucepan, do not use aluminum.
2. Bring ingredients to a slow simmer and reduce by three quarters.
3. Strain sauce before using.

Serving Suggestion: Serve on top of steak and some seafood.

Barbeque Sauce

Courtesy of Chris W. Starkenburg, Harley Marine Services

1	29 ounce can tomato sauce	2	teaspoons Tabasco sauce
1½	cup tomato ketchup	2	teaspoons Liquid Smoke
¾	cup packed brown sugar	½	teaspoon crushed red chilies
½	cup molasses	1	Tablespoon Worcester sauce
1	Tablespoon yellow ground mustard	¼	cup apple cider vinegar
1	Tablespoon Blackfish Seasoning, page 67	¼	cup red wine vinegar
2	teaspoons chili powder		

1. Combine all ingredients in a medium size pot on medium heat until smooth.
2. Reduce heat and simmer for at least 1 hour.

Cooking Suggestion: Brush the sauce on anything you like.

Storing Suggestion: Refrigerate remainder of sauce when finished; will last up to 2 weeks.

Beef Burgundy

Courtesy of Chris W. Starkenburg, Harley Marine Services

Serves: 6-8

2	Tablespoons olive oil	6-7	cups water
1½	Tablespoons garlic, chopped	3-4	Tablespoons beef base or Ajus, page 62
1	cup yellow onion, diced	2	cups mushrooms, sliced
1	cup celery, diced	1	teaspoon fresh ground pepper
1	bay leaf		salt to taste
1½	cups red Burgundy or Merlot wine		**Cornstarch Slurry:**
6	cups precooked beef (prime rib, strip loin, sirloin, beef roast, or tri tip), cubed into 1"x1" pieces	¾	cup cornstarch
		1	cup cold water

1. In large soup pot or cast iron pot add olive oil, garlic, onion, celery and bay leaf. On medium high heat sweat the vegetables until tender.
2. Add the rest of the ingredients except the Cornstarch Slurry.
3. Bring to a boil; reduce the heat and simmer until meat is tender but not falling apart. This will take about an hour depending on the type of meat.
4. Bring back to a slow boil and slowly add the Cornstarch Slurry while stirring a little at a time until you reach desired consistency.
5. Reduce heat.

Serving Suggestion: Serve over pasta, rice or mashed potatoes.

Blackfish Seasoning

Courtesy of Chris W. Starkenburg, Harley Marine Services

3	Tablespoons sweet paprika	1½	teaspoons chili powder
2-3	teaspoons kosher salt or salt substitute	1	teaspoon ground white pepper
1½	teaspoons onion powder	1	teaspoon ground black pepper
2	teaspoons granulated garlic (not garlic salt)	1½	teaspoons whole thyme
1½	teaspoons cayenne pepper	1½	teaspoons whole oregano leaves, rubbed between hands
½	teaspoon crushed red chili peppers	2½	teaspoons dried parsley, rubbed between hands

Combine all ingredients in mixing bowl; you can mix by hand or with electric mixer until well blended.

Large Batch

1	cup sweet paprika	1½	Tablespoons ground white pepper
¼	cup kosher salt or no salt substitute	1½	Tablespoons ground black pepper
¼	cup granulated garlic	¼	cup whole thyme
1½	Tablespoons cayenne pepper (add more to make it hotter)	¼	cup whole oregano, rubbed between hands
1	Tablespoon crushed red chili peppers	¼	cup dried parsley, rubbed between hands
3	Tablespoons chili powder	¼	cup granulated onion

Cajun Candy Rub

Courtesy of Chris W. Starkenburg, Harley Marine Services

2	cups firmly packed brown sugar
½	cup Blackfish Seasoning, page 67
2	teaspoons crushed red chili flakes

Mix in a bowl with electric mixer until blended.

Caramelized Onions

Courtesy of Chris W. Starkenburg, Harley Marine Services

2	Tablespoons butter
2	large sweet or yellow onions, sliced ⅛ inch thick rings
1½	Tablespoons brown sugar
1½	teaspoons balsamic vinegar

1. In a 10" skillet over medium high heat, melt butter and add onion rings.
2. Cook, about 5-8 minutes, then add the brown sugar and balsamic vinegar.
3. Continue to cook until a golden caramelized color.
4. Remove from heat and enjoy with your favorite dishes.

Carrot Fries

1	pound of carrots, cleaned and peeled
2	Tablespoons of vegetable or olive oil
	salt and pepper to taste

1. Preheat oven to 425°F.
2. Grease a cookie sheet with oil and set aside.
3. Cut the carrots into thin strips.
4. Toss carrots with oil and salt in a mixing bowl until well coated.
5. Spread evenly on cookie sheet.
6. Cook until tender, about 20 minutes.

Chris' Cajun Candy Salmon

Courtesy of Chris W. Starkenburg, Harley Marine Services

8	ounce pieces of boneless skinless salmon filet per person
2½	cups Cajun Candy Rub, page 68

1. About 6 hours before serving, coat both sides of filet with Cajun Candy Rub, about 1 Tablespoon per side.
2. Place in freezer lock-top plastic bag or layered on a plate between plastic wrap.
3. Marinate for 5-6 hours in the refrigerator.
4. Place marinated salmon filets on top of foiled baking sheet.
5. Top with 1 Tablespoon Cajun Candy Rub.
6. Place in a preheated oven on high broil with the rack in the middle of the oven.
7. Broil for 6-7 minutes depending on the thickness of filet.
8. Cook the filet until it is a little clear in the middle, once removed from the heat, it will continue to cook.

Blackened Halibut-Salmon-Codfish

Courtesy of Chris W. Starkenburg, Harley Marine Services

6-8	ounce boneless skinless filets of halibut, salmon or codfish per person
	olive oil
	Blackfish Seasoning, page 67
	Butter

1. Lightly brush filets with olive oil.
2. Sprinkle both sides of fish with generous amount of Blackfish Seasoning.
3. Preheat a cast iron skillet on medium high heat.
4. When the skillet is hot, slowly place filet in skillet one at a time.
5. Let sit for 1-3 minutes, then carefully flip and repeat on other side. You are looking for a nice dark crust.
6. Remove the fish when it is a little opaque (clear) in the middle.
7. Place a small piece of butter on top and remove from heat.

Cooking Suggestion: This process is very smoky so you may want to do this out side if you have a BBQ burner.

Serving Suggestion: Serve with Creole Mustard Sauce or nice Mango Chutney.

Bow Tie Pasta Salad with Goat Cheese and Fresh Mozzarella

Courtesy of Chris W. Starkenburg, Harley Marine Services

¼	cup kosher salt	6	mini orange bell peppers
½	cup vegetable oil	1	bunch fresh spinach, cleaned, washed and sliced ½ inch julienne no stems
1	16 ounce box of bow tie pasta	¾	cup sun dried tomatoes, drained from oil and julienne
¼	cup olive oil	12-15	fresh basil leaves, rolled like a cigar and sliced very thin
3	Tablespoons balsamic vinegar	15	small fresh grape cherry tomatoes
1	Tablespoon Dijon mustard	3-4	ounces low fat goat cheese, crumbled
1	teaspoon fresh ground pepper	8	ounces fresh mozzarella, cubed into ¼ inch pieces
1	large sweet onion, peeled and cut into quarters		**Balsamic Vinegar Dijon Sauce:**
10-12	roasted garlic cloves	½	cup olive oil
6	mini red bell peppers	1	tablespoon Dijon mustard
6	mini yellow bell peppers	½	cup balsamic vinegar

1. In a large 5 quart stock pot, add water, salt and vegetable oil and bring to a boil.
2. Add the bow tie pasta, gently stir and reduce the heat so the water is at a slow boil.
3. Cook the pasta al dente this will help keep it from breaking apart.
4. Drain and rinse in cold water until pasta is cooled, after it is cooled and still in the colander, lightly drizzle with olive oil; set aside.
5. In a large bowl, mix olive oil, balsamic vinegar, Dijon mustard and pepper until blended.
6. Then place the onions, garlic cloves and peppers in bowl; roll until everything is coated.

7. Place in preheated BBQ. If possible, use a vegetable grilling screen to keep the vegetables from falling through the grating.
8. Leave remaining liquid in the bowl.
9. Grill vegetables until they have some color and are still al dente; remove from heat and place back in the bowl with liquid and gently top to coat.
10. In Large bowl, place pasta, spinach, tomatoes, basil, roasted vegetables (with extra liquid in bowl, mozzarella and ½ the goat cheese.
11. Gently toss with your hands to evenly blend ingredients.
12. Add salt and pepper to taste.

Serving Sugggestion: Serve in a nice serving bowl and arrange some of the peppers around the edges and top with remaining cheese. Drizzle with Balsamic Vinegar Dijon Sauce and garnish with a large basil sprig. Leave extra sauce on the side in case someone would like to add more.

For better flavor substitute kosher or sea salt for table salt.

Cheese Sauce

Courtesy of Hugo Padilla, Foss Maritime Company

1	8 ounce can cream of chicken soup
½	cup chicken broth
8	ounces or slices of Swiss cheese, mozzarella, or jack cheese, cubbed

1. Bring chicken soup and chicken broth to a boiling point.
2. Add the cheese slowly, stirring constantly until the cheese melts.

If you like to BBQ, eat cruciferous vegetables such as broccoli, cabbage, cauliflower, kale and brussels sprouts. These vegetables contain compounds that activate enzymes in the body that detoxify carcinogenic chemicals called heterocyclic amines (HCAs).

Chicken Caesar Salad

Courtesy of Chris W. Starkenburg, Harley Marine Services

Serves: 6

6	chicken breasts (marinate in caesar dressing for 24 hours before grilling)
1	cup Chris' Caesar Salad Dressing, page 80
4	heads romaine lettuce
	croutons
	parmesan cheese, freshly grated

1. Place chicken breast in a large freezer lock-top plastic bag; and add Chris' Caesar Salad Dressing to the bag.
2. Squeeze excess air out of plastic bag before sealing. Gently squish the breasts around in the bag so that they are evenly coated.
3. Place in the refrigerator for 12-24 hours.
4. Cook the chicken: grill on BBQ, sear in a pan or broil in the oven.
5. After cooking, let the breasts rest 10 minutes before cutting up.
6. Trim off any brown edges from the romaine lettuce.
7. Bunch the romaine together with your hands and gently push down one or two times to flatten. Cut two strips one inch apart from one end to the other.
8. Now rotate 90° while keeping everything bunched together.
9. Cut two more strips one inch apart from one end to the other.
10. Now you will cross cut from side to side one inch apart.
11. This will give you 1"x1" pieces of romaine.
12. Place cut pieces into water and gently toss in the water to clean.
13. Remove from the water and place in a salad spinner to spin off water. If you don't have a salad spinner, place romaine in a colander to let water drip from greens. You can then place in a clean kitchen towel inside a bowl and place into the refrigerator until you are ready to assemble.
14. In a large bowl gently toss romaine with desired amount of dressing and a handful of croutons.
15. Transfer to a plate or large bowl and top with grilled chicken strips.
16. Garnish with a generous amount of parmesan cheese and a lemon wedge.

Storing Suggestion: Prepare lettuce ahead of time and place in freezer lock-top plastic bag lined with a white paper towel, this will last up to 4 days in the refrigerator.

Chicken Crunch

Courtesy of Chris W. Starkenburg, Harley Marine Services

1	10 or 12 ounce box of Cap'n Crunch® cereal
1	10 or 12 ounce box of corn flakes cereal
1	small box of panko bread crumbs (available in the local grocery store)
3	Tablespoons Blackfish Seasoning, page 67
2	cups all purpose flour
6	eggs
1	teaspoon of Tabasco sauce
1	cup of milk
2	pounds of chicken tenders (remove the tendon)
	vegetable oil for deep frying

1. In a blender or food processor blend the Cap'n Crunch® and corn flakes.
2. In a bowl mix blended cereal with panko bread crumbs and 2 Tablespoons Blackfish Seasoning.
3. In a separate bowl mix flour and 1 Tablespoon Blackfish Seasoning.
4. In a third bowl, mix eggs, Tobasco and milk.
5. Add a few chicken tenders at a time, coat well in seasoned flour and shake off excess.
6. Place into the egg mixture making sure that all the floured chicken gets coated well.
7. Place into the cereal mixture, pressing lightly to make sure that you get a good even coat.
8. Place chicken tenders on parchment paper with a little of the breading to keep it dry.
9. Layer on baking sheet using wax paper between to keep chicken pieces from touching.
10. Refrigerate until you are ready to cook.
11. Cooking oil should be heated to at least 325-350°F before placing in the tenders.
12. Cook until golden brown; remove and place on paper towel.
13. Cut one open to make sure they are done. If they need to cook a little longer you can finish cooking in the oven on 350°F for a couple of minutes.

Serving Suggestion: Serve on platter with Creole Mustard Sauce, page 81

Chicken Fried Steak

Courtesy of Chris W. Starkenburg, Harley Marine Services

Serves: 6-8

12	4-5 ounce cube steaks	1	Tablespoon Johnny's Seasoning Salt
3-4	cups good vegetable oil	1	Tablespoon granulated garlic
	Seasoned Flour:	1	Tablespoon onion powder
2	cups flour	1	Tablespoon cracked black pepper
1	cup corn starch	1	teaspoon chili powder
1	cup Yellow corn meal	1	Tablespoon dried parsley, rubbed between hands

1. Mix all dried ingredients in a large bowl.
2. Dredge cube steaks with the Seasoned Flour and lightly press to ensure a good even coat.
3. In a large skillet add 3 cups of vegetable oil and bring the heat up to medium high. (If possible use cast iron skillet, it provides more even heat.)
4. Carefully place one cube steak at a time into the skillet; place the steak into the pan away from you to avoid getting splattered with hot oil. Only add one steak at a time, otherwise you will loose the heat in the oil and the steak will not turn out crispy. You may only get three cube steaks in the pan at a time.
5. Cook equal time on each side until golden brown, about 4-5 minutes per side.
6. Remove from skillet and place on foiled baking sheet until you have all the steaks cooked.
7. Then place in preheated 325°F oven and cook for another 15 minutes.

Skillet Gravy

1	cup yellow onion, diced
1	cup of seasoned flour (above)
2	Tablespoons chicken base
48	ounce of milk (1 quart plus 2 cups)

1. Reduce the heat in the skillet to medium, carefully remove all but ½ cup of the vegetable oil. Leave all the crispy goodies in the pan.
2. Add the onions and cook for about 5 minutes.
3. Sprinkle left over seasoned flour into the pan.
4. Reduce the heat to medium low and continue to stir for 10 minutes, this will cook out the flour taste.
5. Slowly stir in chicken base and milk, until you reach desired thickness, continue stirring until bubbling and smooth. Be careful not to scorch.
6. Remove from the heat, and you are ready to eat.

Serving Suggestion: Serve with garlic mashed potatoes and your favorite vegetable.

Chicken Fried Steak

Chicken Soft Tacos

Courtesy of Chris W. Starkenburg, Harley Marine Services

Serves: 6-8

6	boneless skinless chicken breasts, seared or grilled and sliced on a bias (across the grain)	1	pound cheddar cheese, grated
4-5	Tablespoons vegetable oil	1	pound jack cheese, grated
2	packages taco seasoning	1	bottle favorite salsa
2	fresh garlic cloves, diced fine	16	ounces sour cream
1	medium onion, julienne		**Fixings:**
1	jalapeño pepper, diced fine	2	large tomatoes, diced and drained
1	red bell pepper, julienne	1	bunch of green onions, diced fine
1	yellow bell pepper, julienne	1	head Iceberg lettuce, shredded
12	flour tortillas	2	avocados, sliced and sprinkled with lemon juice and lightly dusted with kosher salt.

1. Cut the center membranes from the chicken breast and place in a bowl.
2. Drizzle breast with 2 Tablespoons vegetable oil and 1 package of taco seasoning; toss until chicken is coated evenly.
3. Sear chicken in a non-stick skillet or grill on BBQ for better flavor.
4. Place the chicken in the refrigerator to cool for about 15 minutes.
5. In a large skillet on medium high heat, place remaining vegetable oil and add garlic, onions and jalapeño; sauté for 3-4 minutes or until onions start to tender.
6. Add the bell peppers and continue to cook for another 3-4 minutes.
7. Add sliced chicken and sprinkle with remaining taco seasoning.
8. Sauté until chicken is warmed through and remove from heat.
9. Place another skillet on medium high heat; when the skillet is hot, place a flour tortilla in skillet, about 30 seconds, flip over and sprinkle with both cheeses.
10. Once the cheese starts to melt, remove and serve.
11. Each person can place desired fixings on top.

Serving Suggestion: Serve with Mexican Red Rice (page, 92) and refried beans.

Chris' Caesar Dressing

Courtesy of Chris W. Starkenburg, Harley Marine Services

1	quart mayonnaise	3	ounces Worcestershire sauce
2	Tablespoons whole marjoram, rubbed between hands	1	Tablespoon Dijon mustard
¾	cup lemon juice (fresh is best)	1	Tablespoon stone ground Dijon
¼	cup red wine vinegar	1	tube (2 ounces) anchovy paste
5	ounce parmesan cheese, grated	2 ½	Tablespoons garlic, chopped fine (Fresh is Best)

Mix well unitl blended and creamy smooth.

Storing Suggestion: Will keep up to 2 weeks in the refrigerator.

Serving Suggestion: Toss with 1"x1" pieces of romaine lettuce and croutons. Top with a squeeze of fresh lemon and grated parmesan cheese.

Citrus Marinade

Courtesy of Chris W. Starkenburg, Harley Marine Services

1	cup olive oil	1	lime, juice
1	Tablespoon sesame seed oil	4-5	large fresh whole basil leaves
2	cloves fresh garlic, smashed	2	teaspoons kosher salt
1	lime, zest (use micro grater)	½	teaspoon fresh cracked pepper
1	lemon, zest (use micro grater)	¼	teaspoon crushed red chilies
2	lemons, juice		

1. In a glass bowl blend all ingredients until smooth.
2. Let marinade set at room temperature for at least one hour before using.

Citrus Scallops on Bed of Avocados

Courtesy of Chris W. Starkenburg, Harley Marine Services

6	large scallops per person (10-15 count scallops)
	Citrus Marinade, page 80
½	avocado per person
	kosher salt and pepper to taste
	fresh herb sprig (for garnish)

1. Marinate scallops in Citrus Marinade for at least 30 minutes; no longer than 1 hour.
2. Remove from marinade and gently place on BBQ grill that has been cleaned and lightly wiped with olive oil or sprayed with vegetable spray.
3. Grill 2-3 minutes on each side or until they are still a little clear in the middle.

Cooking Suggestion: Do not overcook or you will end up with rubbery and chewy scallops.

Serving Suggestion: Place cooked scallops on top of sliced and feathered avocado and drizzle a little of the Citrus Marinade over the top and garnish with herb sprig.

Creole Mustard Sauce

Courtesy of Chris W. Starkenburg, Harley Marine Services

1½	cups mayonnaise	1	Tablespoon Tabasco or favorite hot sauce
¾	cup ketchup	1	teaspoon garlic, chopped
2	Tablespoons Blackfish Seasoning, page 67	2	Tablespoons stone ground Dijon mustard

Mix well in a bowl and keep refrigerated.

Storing Suggestion: Sauce will last weeks in the refrigerator.

Cumin Rice and Beans

Courtesy of Matthew Broomhead, Sea Coast Transportation

Serves: 4-6

1	Tablespoon oil
1	small onion, chopped
2-3	fresh garlic cloves, crushed
1	jalepeño pepper, chopped (optional)
1	Tablespoon cumin
1	cup uncooked rice
1½	cups water
1	12 ounce can black, red or pinto beans, rinsed
	handful of fresh cilantro, coarsely chopped

1. Heat oil in medium saucepan.
2. Sauté onion and garlic (and pepper, if included) until tender.
3. Add cumin and continue sautéing for a couple minutes.
4. Add rice and cook for two minutes, stirring frequently.
5. Add water and bring to a boil. Cover and simmer until water is absorbed, approximately 15-20 minutes.
6. After rice is finished cooking, add rinsed beans, a handful of cilantro, salt and pepper to taste. Heat through and enjoy!

Serving Suggestion: Serve as a side dish with a Mexican meal. Very good when served with a grilled skirt or flank stank, warm flour tortillas or quesadillas and garnished with sliced mangos.

Deep Fried Ravioli

Courtesy of Chris W. Starkenburg, Harley Marine Services

Serves: 6

1	16 ounce package of fresh ravioli uncooked	2	Tablespoons dried parsley
4	cups panko bread crumbs (non-toasted)	1½	teaspoons salt
3	eggs	1	teaspoon black pepper
¾	cup milk		vegetable oil
1½	cups flour	½	Tablespoon Johnny's Seasoning Salt
2	Tablespoons Italian seasoning		parmesan cheese, double grated (for garnish)

1. Blend flour, ½ Italian seasoning, ½ parsley, salt, pepper and Johnny's salt in a container; set aside.
2. In a separate container scramble eggs and milk together; set aside
3. In a third container place panko breadcrumbs, ½ Italian seasoning and ½ parsley; set aside.
4. One piece at a time, place ravioli in egg mixture and lightly coat.
5. Then place egg coated ravioli in flour mixture; shake off excess flour.
6. Coat ravioli in egg mixture again, then lightly press in breadcrumb mixture until coated. If you have any holes dip that end back into the egg mixture and recoat with panko breading.

Panko Bread Crumbs can be purchased at most major grocery stores; however, you can save a lot of money by purchasing at an Asian grocery store.

7. After the ravioli is completely coated, transfer to a baking sheet with a light layer of panko breading, this will help soak up any extra moisture; try not to overlap pieces. Cover with wax paper, then place another light layer of panko, and another layer of breaded ravioli. Do not stack more then 4 layers deep.
8. In Dutch oven or heavy gauged skillet, heat up at least 1" vegetable oil to 325-350°F.
9. Carefully place 3-4 pieces with tongs into the oil away from you.
10. Cook until light golden brown and place on cookie rack on top of paper towel.
11. Lightly salt and dust with a good layer of parmesan cheese right after it comes out of the fryer.

Food Preparation Suggestion: When doing the breading procedure use one hand for the wet part of the breading and the other hand for the dry part of the breading.

Cooking Suggestion: Do not over load the fryer, the oil will lose heat and make your food come out greasy.

Serving Suggestion: Serve with both Pesto (page 95) and a spicy red sauce

Storing Suggestion: Prepare the coated raviolis ahead of time. To prevent sogginess, do not store in the refrigerator for more than 2 days. The coated raviolis freeze well; just slightly thaw before cooking.

Fresh Grilled Vegetables

Courtesy of Chris W. Starkenburg, Harley Marine Services

1	green zucchini, sliced 3/8 inch thick on a bias (across the grain)
1	yellow zucchini, sliced 3/8 inch thick on a bias
1	yellow sweet onion, quartered and separated
6-8	medium size brown or white mushrooms
6-8	fresh shitake mushrooms
6-8	fresh asparagus tips
3	whole garlic cloves, smashed
1/3	cup olive oil
1-2	teaspoons salt
7-8	cranks of fresh pepper
	Oil Mixture:
1	Tablespoon balsamic vinegar
1	teaspoon sesame seed oil

1. In large stainless steel or glass bowl, place cleaned and cut vegetables.
2. In a separate bowl, whisk together olive oil and the other ingredients.
3. Drizzle oil mixture over the vegetables, and gently toss with your hand to ensure all vegetables get coated.
4. Cover and refrigerate, about 30 minutes.
5. When your grill is good and hot, place vegetables on grill with tongs, let the vegetables cook to your desired tenderness and remove from heat and place on a serving platter.
6. Lightly drizzle with oil mixture and serve.

Frijoles (Beans)

Courtesy of Matthew Broomhead, Sea Coast Transportation

2	slices bacon or ham
2	tablespoons olive oil
1	clove garlic, minced
½	onion, diced
½	cup celery, diced
¾	16 ounce can tomato sauce
½	cup water
1	teaspoon white or apple cider vinegar
2	16 ounce cans red or kidney beans
½	green pepper, cubed
	salt and pepper to taste
	fresh cilantro, coarsely chopped

1. Brown bacon or ham in large sauce pan over medium high to high heat.
2. Add oil to pan and heat; sauté garlic and onions then celery.
3. Add tomato sauce, water, vinegar, beans and green pepper.
4. Salt and pepper to taste.
5. Stir, cover and simmer on medium heat, about 20 minutes.
6. Before serving, stir in small handful of cilantro.

Garlic Mashed Potatoes

Courtesy of Chris W. Starkenburg, Harley Marine Services

2-3	pounds baby red or yellow new potatoes
½	cup butter
½-¾	cup parmesan or romano cheese, grated
1-2	cups half-and-half
6	cloves of fresh roasted garlic, smashed and diced
	kosher salt and pepper to taste

1. Place potatoes in a large pot and cover with water, bring to a boil.
2. Reduce heat to a slow boil and cook until they are fork tender.
3. Remove from the heat and let sit for 10 minutes.
4. Drain off water, add butter, cheese and ½ the half-and-half.
5. Mash with potato masher or use hand mixter and add more of the half-and-half to desired thickness. Do not over mix, it is better with some lumps
6. Salt and pepper to taste.

Goya Seasoning

Courtesy of Matthew Broomhead, Sea Coast Transportation

1	Tablespoon garlic salt
1½	Tablespoons dried oregano
1	Tablespoon turmeric
1	Tablespoon salt
1½	Tablespoons pepper
1	Tablespoon cumin

Mix all ingredients together.

Jasmine Rice

Courtesy of Matthew Broomhead, Sea Coast Transportation

1	cup jasmine long grain rice
1½	cups water
1	teaspoon salt
2	teaspoons olive oil

1. Bring water, salt, and oil to boil in large sauce pan.
2. Add rice and cook until almost dry.
3. Cover and lower heat to simmer for 20 minutes.

Fried Rice is a great way to use up leftover vegetables. See Pork Fried Rice, page 97

Macadamia Nut Crusted Halibut

Courtesy of Chris W. Starkenburg, Harley Marine Services

Serves: 6

6	8 ounces boneless skinless halibut filets
1	cup flour
2	teaspoons Johnny's Seasoning Salt
2	eggs
½	cup milk
2	cups panko bread crumbs
½	cups toasted macadamia nuts, chopped fine to medium
1-2	cups vegetable or olive oil

1. In shallow container, mix flour and Johnny's salt; set aside.
2. In a second shallow container scramble eggs and milk together; set aside.
3. In a third container, mix bread crumbs and macadamia nuts together; set aside.
4. Take each filet, one at a time, and set down just coat or crust the top of each filet in the flour; do not coat the entire filet.
5. Pick up and coat the same side in the egg mixture.
6. Then pick up and coat the same side in the bread crumb mixture.
7. Place each filet on a baking dish lightly sprinkled with bread crumb mixture, coated side down and refrigerate until you are ready to cook.
8. In a none stick skillet, place ¼ inch of oil in pan and heat to medium high heat. Carefully place filet bread crumb side down and lightly brown, about 1-2 minutes.
9. Carefully remove and place crusted side up on lightly olive oiled baking dish.
10. Place in preheated oven on 375°F for about 4-5 minutes or until lightly clear in the middle.

Serving Suggestion: Serve with mango chutney or a citrus aioli sauce.

Macadamia Nut Crusted Halibut, page 89

Meatballs

Courtesy of Chris W. Starkenburg, Harley Marine Services

1	pound bulk pork breakfast sausage	1	teaspoon red chilies, crushed
1	pound ground beef (ground chuck recommended for flavor)	1	Tablespoon kosher salt
½	large sweet onion, finely chopped	10	grinds fresh black pepper or to taste
6	medium cloves of garlic, smashed and chopped fine	1	cup seasoned bread crumbs
¼	cup fresh parsley, chopped	1	Tablespoon Worcestershire sauce
2	teaspoons dried oregano, crumbled	2	eggs, lightly beaten with a fork
1	Tablespoon fresh basil	2-4	cups vegetable oil

1. In a large bowl place all the ingredient, except the vegetable oil.
2. Mix thoroughly with hands until everything is blended together.
3. Portion meatball into desired size.
4. In a heavy gauged pan, add the vegetable oil and place on medium high heat.
5. When oil is hot, carefully add meatballs one at a time.
6. The meatballs should be covered in oil like you are deep-frying.
7. Cook for about 3-5 minutes or until the meatballs are almost cooked through.
8. Remove from the oil and place on cooling rack to remove any excess oil.
9. Once they have cooled, you can place in your favorite sauce.

Store jars of quality marinara and packages of dried pasta for a quick and satisfying meal.

~

Is your marinara sauce too acidic – try a pinch of baking soda to neutralize the acid

Mexican Red Rice

Courtesy of Chris W. Starkenburg, Harley Marine Services

2	Tablespoons butter
2	cups rice (short grain works best)
3	teaspoons chicken base (or 2¼ cups chicken broth and omit water below)
2¼	cups hot water
2	Tablespoons yellow onion, diced
1	garlic clove, diced
1	cup tomatoes, diced

1. Use a cast iron skillet, Dutch oven or heavy pot that can be placed in the oven.
2. On medium heat, place butter and rice in pan; stir until rice is golden brown.
3. In a blender, place chicken base, water or chicken broth, onion, garlic, and diced tomatoes.
4. Blend until smooth, then pour in pot with the rice.
5. Stir until it comes up to a boil, then cover and place in a preheated 350°F oven.
6. Cook in oven for about 20-25 minutes. You can also finish on stovetop the standard way.

Orzo Pasta with Shitake Mushrooms

Courtesy of Chris W. Starkenburg, Harley Marine Services

½	cup vegetable oil
2	Tablespoons kosher salt
1	box of orzo pasta
	olive oil
1½	teaspoons sesame seed oil
1	small sweet onion, diced
2	cloves garlic, diced
½	cup sweet red bell pepper, diced
1½	cups shitake mushrooms, sliced
2-3	Tablespoons fresh cilantro or Italian parsley
	salt to taste

1. In a large pot, bring water, vegetable oil and salt to a full boil.
2. Add orzo pasta and stir to keep from sticking to bottom.
3. Cook until just tender but do not over cook.
4. Drain in colander and rinse in cold water until orzo is cold and the cooking has stopped.
5. While the orzo is still in colander, drizzle with olive oil to keep it from sticking.
6. In large non stick sauté pan, add sesame seed oil.
7. On medium to medium high heat, sweat onion until tender, then add garlic and red bell peppers and shitake mushrooms.
8. Add the orzo pasta and fresh herbs of choice.
9. Sauté for 7-8 minutes or until hot and serve.

Oven Broiled Salmon with Pesto

Courtesy of Chris W. Starkenburg, Harley Marine Services

1	boneless skinless salmon filet
	olive oil
	fresh ground black pepper
1-2	cups fresh Pesto Sauce , page 95
	fresh parmesan cheese, grated
	sun dried tomatoes (for garnish)

1. Preheat oven to 400°F.
2. On foiled baking sheet or baking dish, lightly oil and place salmon filet.
3. Lightly rub the top of salmon filet with olive oil and lightly crank some fresh ground pepper on top.
4. Just before placing salmon in the oven, switch temperature to high broil and make sure the oven rack is 6 inches from the broiler.
5. Cook the salmon 3-5 minutes or until internal temp is 120°F or slightly clear (not cooked all the way through) in the middle. The fish will continue to cook after it is removed from the heat.
6. Remove from oven and place a generous amount of fresh Pesto Sauce on top of salmon filet.
7. Dust with some freshly grated parmesan cheese and return to oven for another minute or two.
8. Remove from oven and place on serving dish and garnish with diced sun dried tomatoes.

Pesto Sauce

Courtesy of Chris W. Starkenburg, Harley Marine Services

4	ounce package fresh basil leaves de-stemmed
1	cup parmesan or romano cheese or a combination of each, grated
1½	cups olive oil
½	teaspoon fresh ground black pepper
1	teaspoon kosher salt
2	teaspoons fresh lemon juice
6	fresh garlic cloves
½	cup toasted pine nuts

In electric blender, add all ingredients and mix until smooth. You may have to push the ingredients to the bottom of blender several times to get it started.

Serving Suggestion: Add Pesto Sauce to pasta or brush over charbroiled seafood after you flip it.

Storing Suggestion: Freeze until future use.

Pan Seared Halibut w/ Lemon Curry Sauce

Courtesy of Mike Wentworth, Western Towboat Company

1	filet of halibut
3	Tablespoons olive oil
2	Tablespoons flour
1	Tablespoon curry powder
1	lemon, juice
1	cup heavy cream
1	teaspoon lemon zest
	fresh dill (for garnish)

1. Preheat oven to 400°F.
2. Cut fillet into 2 large pan sized pieces.
3. Heat oil in large pan until lightly smoking.
4. Sear each side of fish until lightly browned.
5. Place fish on foil lined baking sheet and put in oven; cook fish until 140°F.
6. Save oil in pan, reduce heat and add flour. In the following order, add curry powder, lemon juice, cream and lemon zest.
7. Remove from heat and stir until thickened.
8. Place fish on serving platter, cover with sauce and sprinkle with fresh dill.
9. Serve hot.

Leftover Suggestion: Use leftover halibut for Halibut or White Fish Chowder (page 56).

Pork Fried Rice

Courtesy of Chris W. Starkenburg, Harley Marine Services

Serves: 8-10

3	cups long grain rice
4½	cups water
1	teaspoon sesame seed oil
8	strips of smoked bacon, diced
6	large eggs, beaten like scrambled eggs
1	Tablespoon sesame seed oil
⅓	cup plus 2 Tablespoons Soy sauce
1½	bunches fresh green onions, finely diced

1. Rinse and drain rice 2 times before adding water and sesame oil into your automatic rice cooker. After rice is cooked open the lid and let the rice cool completely. You can cook the rice in the morning or use left over rice.
2. Heat a large non-stick pan or wok to medium high heat.
3. When the pan or wok is hot, add the diced bacon and continue to stir until crisp.
4. Remove bacon from pan, place in glass dish and leave 2 tablespoons of bacon grease in pan.
5. Place pan back on stove and bring the heat up to medium.
6. Gently pour in scrambled eggs and stir until cooked.
7. Place eggs in the same dish as the cooked bacon.
8. Add 1 Tablespoon sesame seed oil, plus remaining bacon grease into pan and place back on medium high heat.
9. Slowly add the cooled rice while stirring to break up any lumps.
10. Cook and continue to stir for another 5 minutes.
11. Evenly pour soy sauce over the rice and stir until blended.
12. Add the bacon, eggs and ½ of the diced green onions.
13. Place back into the rice cooker to keep warm until you are ready to serve.
14. Continue to stir until well mixed.
15. Garnish with remaining green onions and serve.

Pork Fried Rice, page 97

Seafood Stew

Courtesy of Mike Cahill, Dunlap Towing Company

½	pound bacon	2	cans tomato sauce
⅛	cup olive oil	2	cans tomato paste
1	bay leaf	1	can diced tomatoes
½	cup Blackened Redfish Magic®	1	can whole kernel corn
½	onion, chopped	2-3	red potatoes diced (cook in microwave for 6 minutes)
½	green pepper	1	pound prawns
6	garlic cloves	1	pound scallops
2	jalapeño peppers, chopped	2	pounds cod

1. In a Dutch oven, fry bacon and drain.
2. Add olive oil, bay leaf, Blackened Redfish Magic®, onion, green pepper, garlic and jalapeño peppers.
3. Sauté until carmalized.
4. Add tomato paste, tomatoes and corn.
5. Mix and heat.
6. Add potatoes.
7. Add seafood.
8. Slow heat until seafood is cooked.

Seared Caesar Halibut Cheek on Kaiser

Courtesy of Chris W. Starkenburg, Harley Marine Services

Serves: 1

1	6 ounce halibut cheek or halibut filet
3	ounces Chris' Caesar Dressing, page 80
	fresh baked kaiser roll
2	Tablespoons olive oil
3	Tablespoons fresh parmesan cheese, grated
1	ounce romaine lettuce, cut like angel hair
	purple onion rings
	lemon wedge

1. Marinate halibut cheek or filet in 1 ounce Caesar dressing for about 1-3 hours.
2. Slice kaiser roll in half and drizzle with olive oil; dust with fresh parmesan cheese; and, grill in none stick pan until golden brown.
3. In the same pan turn up the heat and sear the halibut for about 1-2 minutes per side. It is very important not to over cook; lightly break open the halibut filet, it should be a little clear in the center.
4. Remove from heat and the pan.
5. Lightly toss the romaine with the dressing.
6. Lightly spoon dressing onto both sides of the kaiser roll and place halibut filet on bottom half and garnish with fresh parmesan.
7. On the top half of the kaiser roll, place the romaine and purple onion rings.
8. Garnish with lemon wedge.

Serving Suggestion: Serve with Carrot Fries.

Spicy Apple Pork Loin Roast

Courtesy of Chris W. Starkenburg, Harley Marine Services

Serves: 8-10

6-7	pounds boneless pork loin
¼	cup olive oil
1	Tablespoon kosher salt
½	Tablespoon fresh cracked pepper
1½	teaspoon ground cayenne pepper
1	12 ounce squeeze bottle of honey Dijon mustard
⅔	cup brown sugar
2	Fuji apples, peeled and cored, cut in half and slice very thin in ½ moon shape. Place in lemon water (¼ cup lemon juice mixed with ¾ cup water) to keep from turning brown.
⅓	cup honey

1. On foiled or a non-Stick baking sheet, lightly coat with olive oil.
2. Lightly dust pork loin on all sides with salt and pepper.
3. Place loin fat side up and lightly sprinkle with cayenne pepper.
4. Squeeze a generous coating of honey Dijon mustard on top of the pork loin.
5. Sprinkle with ⅓ cup brown sugar.
6. Gently place drained apple slices from each end of the pork loin to the center (Like fish scales) then make small circle of apples in the middle to completely cover the pork loin.
7. Squeeze more of the mustard on top.
8. Drizzle ⅓ cup of honey evenly over the top.
9. Sprinkle with the remaining ⅓ cup of brown sugar.

10. Place pork loin in a preheated 400°F oven on the middle rack for 20-25 minutes, this will sear the pork loin.
11. After 20-25 minutes, reduce the heat to 300°F and do not open the oven door.
12. After another 25 minutes, carefully baste the pork loin with the pan juices and sprinkle a little more brown sugar on top; continue to bake for approximately 25 more minutes or until the internal temperature reaches 155°-160°F (total cook time is approximately 1½ hours).
13. Remove from the oven and let stand for at least 10-12 minutes before slicing.

Steak Marinade

Courtesy of Chris W. Starkenburg, Harley Marine Services

1	lemon, zest (use micro grater) and juice
½	cup olive oil
1	Tablespoon fresh ground pepper
4-6	garlic cloves, smashed flat with knife and chopped
4	teaspoons white sugar
¾	cup Worcestershire sauce
2	Tablespoons Dijon mustard
⅓	cup balsamic vinegar
3	green onions, cut into 2 inch pieces

1. Combine all ingredients.
2. Place Steaks in a large freezer lock-top plastic bag or a marinating pan.
3. Pour marinade over steaks and marinade overnight in the refrigerator.

Always marinate food in the refrigerator.

Stuffed Chicken Breast A La Hugo

Courtesy of Hugo Padilla, Foss Maritime Company

8	boneless chicken breasts
8	slices of ham
2	cups cheddar cheese, shredded
8	cups of Stuffing, page 105
	salt and pepper to taste
	Cheese Sauce, page 74
1	small jar pimentos
	paprika
	fresh parsley

1. Preheat oven to 350°F.
2. Place the chicken breast on a cutting board; remove the skin and excessive fat.
3. Pound each chicken breast flat with a mallet. (Place plastic wrap over chicken or place chicken in freezer lock-top plastic bag prior to pounding.)
4. Place a ham slice, cheese and a large spoonful of stuffing on top of each chicken breast.
5. Roll the chicken breast away from you making sure nothing falls out of the wrap.
6. Sprinkle salt and pepper, do the same thing with the other chicken breasts.
7. Place rolled chicken breasts (seam down) next to each other in a greased baking pan (13"x9"x2").
8. Bake for 45 minuntes.
9. While chicken is baking, make Cheese Sauce.
10. Take out of the oven and pour cheese sauce over each piece of chicken.
11. Place a few pimentos on each piece.
12. Sprinkle with fresh parsley and paprika for garnish.

Serving Suggestion: Serve with rice pilaf and asparagus.

Stuffing

Courtesy of Hugo Padilla, Foss Maritime Company

10	slices of bread, cut into small pieces
	garlic salt
	pepper
	olive oil
½	cup onion, chopped
2	celery stalks, chopped
2	garlic cloves
	butter
1	cup chicken broth

1. Toss bread with a sprinkling of garlic salt, pepper and olive oil.
2. Spread on a greased cookie sheet.
3. Bake for 15-20 minutes at 350°F until toasted.
4. In a Dutch oven, sauté onion, celery and garlic in butter until onions are translucent.
5. Add chicken broth.
6. Add toasted bread.
7. Let mixture cool for 5 minutes.

Teriyaki Dipping Sauce

Courtesy of Chris W. Starkenburg, Harley Marine Services

2	cups water
1	cup firmly packed brown sugar
1½	cups white sugar
2	Tablespoons fresh ginger, chopped
1½	Tablespoons fresh garlic, chopped
1	teaspoon crushed red chilies
1½	Tablespoons sesame oil
1½	cups soy sauce
	Cornstarch Slurry:
2½	Tablespoons corn starch
¼	cup cold water

1. In sauce pan, bring water, brown and white sugar to a slow boil until all the sugar is dissolved.
2. Add all other ingredients, except the Cornstarch Slurry.
3. After the mixture has simmered for 30 minutes, remove from stove and strain to remove all of the solids.
4. Put liquid back into pan and bring to a slow boil.
5. While stirring with whisk, slowly add Cornstarch Slurry.
6. Cook until sauce returns to a boil. If you wish to make a thicker sauce, add more slurry. (Note: Sauce will thicken after refrigeration.)
7. Remove from heat.
8. Refrigerate any unused sauce.

Teriyaki Marinade

Courtesy of Chris W. Starkenburg, Harley Marine Services

2	cups water
1	cup firmly packed brown sugar
1½	cups white sugar
2	Tablespoons fresh ginger, chopped
1½	Tablespoons fresh garlic, chopped
1	teaspoon crushed red chilies
1½	Tablespoons sesame oil
1½	cups soy sauce

1. In sauce pan, bring water, brown and white sugar to a slow boil until all the sugar is dissolved.
2. Add all other ingredients.
3. After mixture has simmered for 30 minutes, remove from heat and let cool.
4. After mixture has cooled, place in desired container with airtight lid and store in refrigerator.
5. After marinade has fully cooled, it is ready to use.

Recommended Marinating Times:
Seafood 4-8 hours
Chicken 8-24 hours
Beef 24-48 hours

Storing Suggestion: Marinade will last in the refrigerator for about 2 weeks.

Never use the same liquid that you have marinated in for a glaze or dipping sauce.

Cream Cheese Frosting

Courtesy of Chris W. Starkenburg, Harley Marine Services

8	ounces cream cheese
½	cup butter, softened
2½	cups powdered sugar
1	teaspoon pure vanilla

Combine all ingredients in a bowl until creamy.

Substitute maple syrup or honey for sugar in recipes

Ginger Spice Cookies

Courtesy of Bruce Hamar, Dunlap Towing Company

1½	cups sugar
2	cups flour
1	teaspoon baking soda
1	teaspoon ground cinnamon
1	teaspoon ground cloves
1	teaspoon ground ginger
⅛	teaspoon kosher salt
¾	cup vegetable shortening at room temperature
1	egg
¼	cup molasses

1. Preheat oven to 375°F.
2. Line a large baking sheet with parchment paper.
3. Place ½ cup of sugar in a small bowl for rolling balls of dough.
4. In a medium bowl, combine the flour, baking soda, cinnamon, cloves, ginger, and salt and whisk until no streaks appear; set aside.
5. Using a mixer on medium high speed, combine the shortening and the remaining sugar; mixing until light and fluffy, about 4 minutes.
6. Add the egg and molasses, incorporating to combine.
7. Reduce the speed to low and gradually add the dry ingredients; mixing until dough is formed.
8. Using your hands, shape the dough into golf ball size portions.(about 2 tablespoons).
9. Roll the balls in sugar and place half of them 2 inches apart on the prepared baking sheet.
10. Bake for 12 minutes. (The tops will be cracked; the insides should appear slightly underdone.)
11. Repeat with the remaining dough.
12. Let the cookies sit on baking sheet for 5 minutes before transferring them to a wire rack to cool completely.

Lemon Bars

Courtesy of Chris W. Starkenburg, Harley Marine Services

1	16 ounce box angel food cake mix
1	22 ounce can lemon pie filling
1	cup finely shredded coconut
1	lemon, zest (use micro grater)
	Cream Cheese Frosting, page 108 (add the zest from 1 lemon)
1	bag of lemon drops (for garnish)

1. Preheat oven to 350°F
2. In mixing bowl, combine cake mix, pie filling, and coconut.
3. Stir until thoroughly mixed. (Do not add any other liquids.)
4. Spread this mixture in a 10"x15" baking pan.
5. Bake for 30 minutes.
6. Let cool before frosting.
7. Frost the top with Cream Cheese Frosting and refrigerate overnight.
8. Cut into 1½"x1½" squares and garnish with lemon drops.

Plantains (Frying Bananas)

Courtesy of Matthew Broomhead, Sea Coast Transportation

firm yellowish-green plantains (more green)
vegetable oil
salt to taste

1. Remove skin (all traces).
2. Slice diagonally into pieces, not more than ¼ inch thick.
3. Fry in a small amount of vegetable oil on medium heat until brown.
4. Remove and salt to taste.
5. Serve warm.

Cooking Suggestion: For crispy plantains, strain cooked plantains on a piece of paper towel. Smash semi-cooked plantains with a fork. Fry again on medium high heat until crispy.

My Galley Favorites

Write your favorite healthy and hearty recipes on the following pages.

Recipe Title:	

Recipe Title:	

Recipe Title:	

Recipe Title:	

Recipe Title:	

Recipe Title:	

Recipe Title:	

Recipe Title:	

Recipe Title:	

Recipe Title:	

Recipe Title:	

Recipe Title:	

Recipe Index

Resources and References

Andrew Weil, M.D., http://www.drweil.com. Dr. Weil's health advisor website is an excellent resource for healthy aging and nutrition.

Pirello, Christina. *Cooking the Whole Foods Way: Your Complete, Everyday Guide to Healthy, Delicious Eating with 500 Recipes, Menus, Techniques, Meal Planning, Buying Tips, Wit & Wisdom*, March 1997.

The American Waterways Operators. *Responsible Carrier Program 2006*, October 2006.

United States Coast Guard. *Crew Endurance Management Practices: A Guide for Maritime Operations*, January 2003.

United States Coast Guard. *Crew Endurance Management Practices: A Guide for Maritime Operations Addendum*, September 2005.

United States Department of Agriculture. *Safe Food Handling: Basics for Handling Food Safely*, September 2006.

ACKNOWLEDGEMENTS

The *Galley Chef* is the result of companies collectively participating to assist inexperienced cooks in the galley and provide a refresher for the more experienced "Galley Chef", while incorporating industry standards and regulations.

Thank you to everyone who submitted recipes and ideas especially to: Captain Jeff and Liz Slesinger; Captain Russ Johnson; Sara Mazurek; Deborah Franco; Captain Mike Curry; Stephanie Wright; and, Matthew Broomhead.

We would like to thank Daniel "Cory" Bilton and the crewmembers of Western Towboat Company for their generous contribution toward creating this book.

A special thank you to Captain Chris Starkenburg, of Harley Marine Services, Inc. for providing most of the delicious recipes and meal planning suggestions; you are truly one of the top "Galley Chefs" in the maritime industry.

A portion of the profits from this book will be donated to the Shriners Hospitals for Children -Portland, Oregon.

Wishing you Safe Voyages,
Dean and Dione Lee